Ice has a huge influence on our daily weather, our climate, and our landscape.

Ice also appears to us in myriad strange and wonderful forms, from film frost to needle ice, from cat ice to icicles, and from hair ice to snow.

In *The Story of Ice*, find out how ice can do all this as you learn important scientific concepts and gain an appreciation for this amazing crystal. Written by an ice and snow scientist and an accomplished storyteller, *The Story of Ice* follows the lead author's very popular earlier book *The Story of Snow*.

Acclaim for *The Story of Snow*:

❄ "The sort of riveting exhibition that will have eyes locked to the pages."
  - *The Bulletin of the Center for Children's Books*, starred review

❄ "With never a hint of hyperbole, the authors communicate such a contagious sense of wonder that few readers will be able to resist."
  -*Booklist*, starred review

❄ NSTA Outstanding Science Trade Book for Students K-12
❄ American Meteorological Society Louis J. Battan Author's Award winner
❄ Bulletin of the Center for Children's Books Blue Ribbon winner
❄ American Association for the Advancement of Science/Subaru SB&F Prize finalist
❄ New York Times 100 Best Titles for Reading and Sharing
❄ Chicago Public Library's Best of the Best

About the characters of this story...

Our central character:

ICE is solid water, a crystal made of water molecules that melts at 32 °F (0 °C).

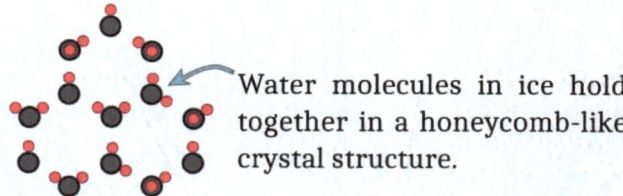

Water molecules in ice hold together in a honeycomb-like crystal structure.

Ice can take on many distinct shapes and sizes, each with their own name. So, just as a person can be a farmer, a nurse, a teacher, or many other things, ice can be snow, icicle, ice cube, graupel, rime, glacier, or many other things. They are all solid water, though they can also contain other things such as air, liquid, and debris.

Our two supporting characters:

WATER is melted ice, a wet liquid of water molecules that flows easily.

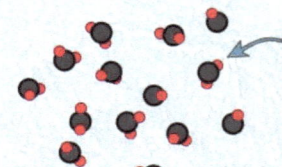

Water molecules in water stay close together, but do not have a rigid structure like ice.

Water can be in the form of a drink, a raindrop, a puddle, a lake, or many other things. As a small droplet, water can remain liquid upon cooling down to -40 °F (-40 °C).

VAPOR is an invisible gas of water molecules, usually mixed in the air.

Water molecules in vapor are spaced far apart (compared to that in water). They zip around at the speed of sound, often colliding and changing direction.

These characters often **change** into each other:

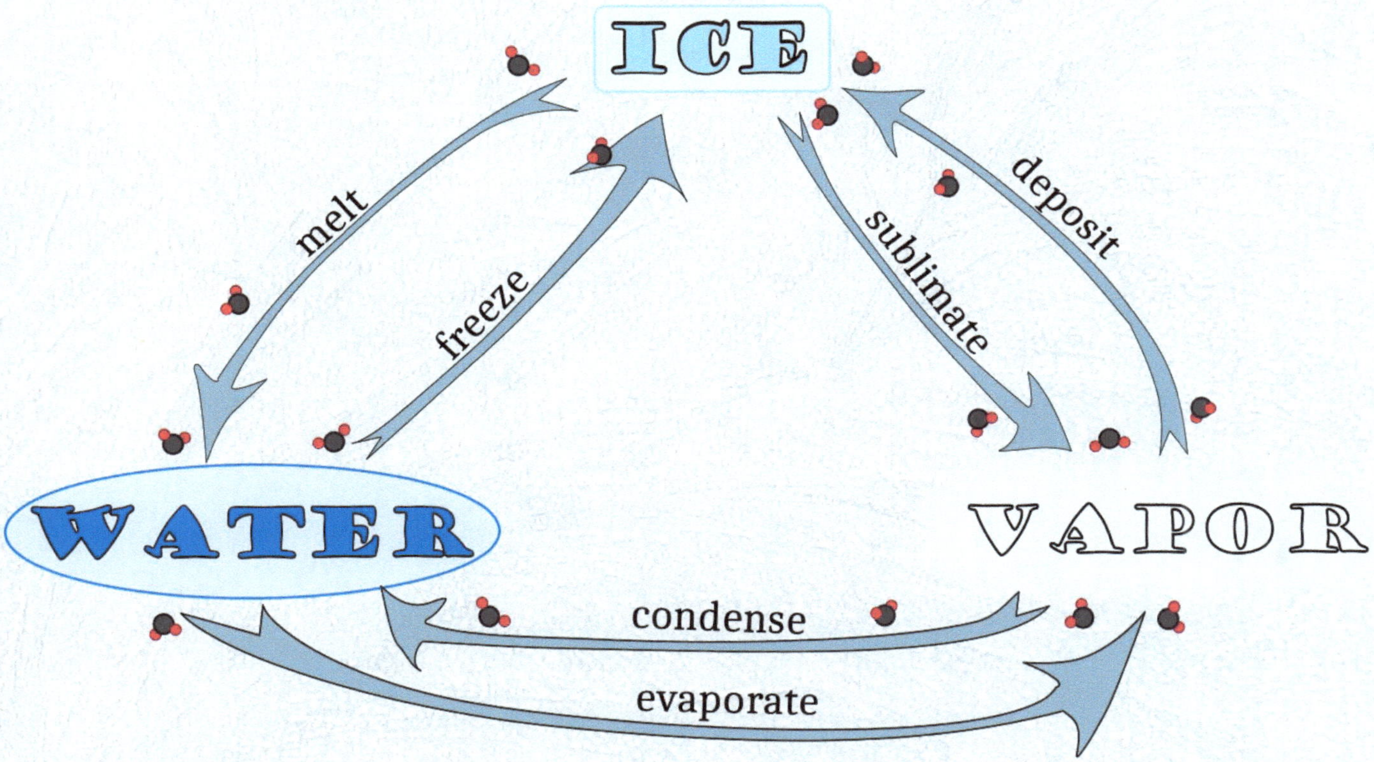

ICE

melt
freeze
sublimate
deposit

WATER

condense
evaporate

VAPOR

For example, in the case of snow, vapor deposits directly into ice.

But the most important character in this story is YOU.

Ice topics with checklist

	Pages	Observed?
Needle ice	10-11	☐
Cat ice (crunchy puddle)	12-15	☐
Black ice	15	☐
Pond ice	17	☐
Hair ice	18	☐
Ribbon ice	18	☐
Hoar frost, columnar	19	☐
Hoar frost, tabular	19	☐
Icicle, growing (hollow tip)	20-21	☐
Ice stalagmite	20	☐
Rime	22-23	☐
Film frost	24-25	☐
Rain	26-29	☐
Snow, dendrite (star)	27,36,37	☐
Snow, column or needle	27,37	☐
Snow, rimed	27,37	☐
Graupel	27	☐
Thunder and lightning	30-33	☐
Hail	33	☐
CZ arc	34-35	☐
Firn and glacial ice	38	☐
Ice core	39	☐
Glacier	40	☐
Iceberg	41-42	☐
Ice ages	45	☐
Fun facts	49	☐
Parent/teacher guide	50-54	☐

*I had the sudden feeling of being drawn
upward through the slowly falling snow.*
— Imaizumi Masato

For those willing to look closer
-JN

For Robin and Asaki
-SN

Text, illustrations, and photos copyright©2025 by Jon and Sam Nelson. All rights reserved. No part of this book may be reproduced in any form without written permission of the authors.

ISBN: 979-8-9938762-0-7

All photos by Jon Nelson except:
Needle ice on left side of page 10 by Sam Nelson
Cat ice photo, middle of page 12 by Kris Buzard
Ribbon ice on page 18 and back cover by Brian Swanson
Snow and ice particles on pages 27 and 37 from the Magono-Lee collection
Snow crystal photos pages 37, 43, and back cover by Mark Cassino
Ice core on page 39 from Wikipedia Commons
Frozen river foam on back cover by Karen Healy

Drawings by Jon Nelson
Background patterns on some pages (e.g., 1,4,6) from Jon's photos of film frost.
Front cover art by Alaina Elsie, designed & modified by the authors.

For more information on ice and related topics, including additional activity suggestions, check for updates on the author's blog at
https://www.wonderintheair.com

The Story of Ice

Exploring Weather, Chemistry, & Physics with Nature's Most Common Crystal

By Jon Nelson and Sam Nelson

If ice could tell a story,

what would it be?

Ice makes the ground crunchy.

Hear that crunchy ground underfoot? You are smashing needle ice.

crunch! crunch! crunch!!

In soil, needle ice often pushes up rocks and twigs.

Needle ice can also grow from pavement ...

... as well as from the tops of small pebbles..

Needle ice draws water from the ground.

Needle ice grows upward from the ground, drawing water from below. As this water squeezes through narrow gaps in the soil to freeze, the resulting ice becomes narrow columns (like tiny skyscrapers).

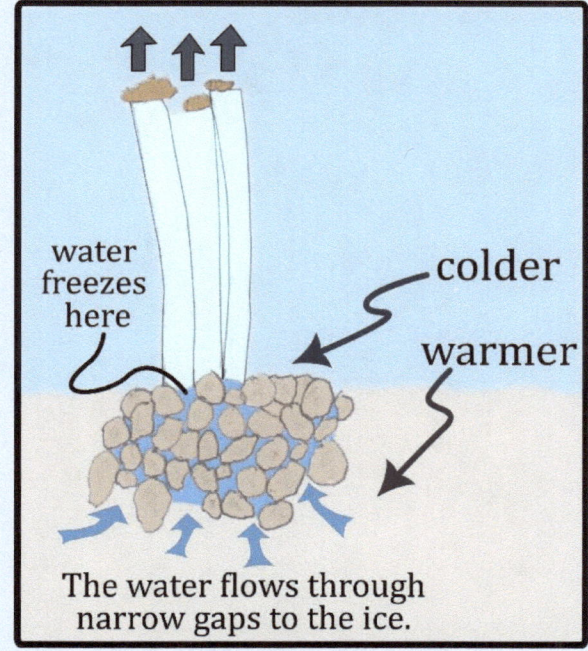

When you step on these columns, they break, making the crunchy sound you hear.

Needle ice on the ground.

As the ground freezes, dirt, rocks, and twigs get pushed up by needle ice. The power of growing ice is so strong that it can move building-size boulders or crack big rocks apart.

11

Near the needle ice—a frozen puddle.

Ice forms over puddles.

I'll smash it.

Won't the water splash?

Ice on a puddle can show straight lines,...

...circular curves,...

...and both curved and straight lines.

Wait, there's no splash!?

There's no water under this ice.

Where did the water go?

The culprit of all this is .... that needle ice!

As needle ice grows, it collects groundwater.

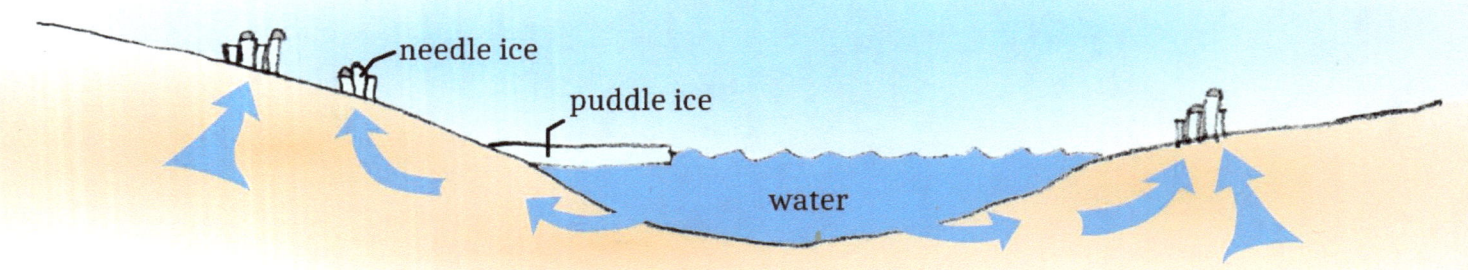

Puddle ice grows quickly across the top. But if needle ice is nearby, it also grows by collecting water from the ground **and** from the puddle, emptying the puddle's water.

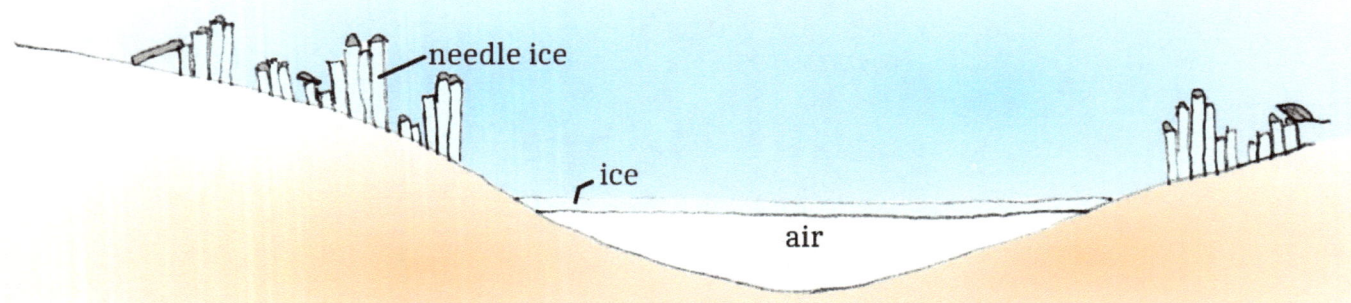

Later, the needle ice may melt and refill the puddle with water. So, if you see the puddle ice one day, come back the next day—it may appear again!

The ice over such a "hollow" puddle is sometimes called *cat ice*.

Cat ice looks white because its top and bottom sides are bumpy and air is below — like a plate of frosted glass.

water under the ice

once was water's edge

Cat ice is the white and thin ice layer over a hollow puddle. The name comes from the idea that only cats can walk on it without breaking through.

Where water touches the ice, the ice looks darker. The thick curvy lines in the white cat ice are places where the edge of the water once made contact.

If the puddle was shallow, all of the water may freeze, leaving no air gap. Such ice may look black. So it is sometimes called "black ice".

How does a puddle freeze?

Ice forms first where the water is coldest.

Can you guess where?

Ice starts from the top and at the shore.

⇩

Then, ice spreads to cover the top. At first, it is very thin and clear.

Clear "blades" of ice taken from a puddle as it was freezing across the top. The blades are very thin. Because of their fern-like shape, we call them "dendrites".

Ice on a pond or lake forms like that on a puddle.

Unlike a puddle, the water level hardly drops.

Yet the ice can still make puzzling patterns.

During freezing of a pond, the water level may go down, but plenty of water remains.

The cat ice at the shore shows that the ice began there, but the water level dropped before the ice spread completely across.

Ice can also begin from objects that stick out of the water.

If the weather stays cold, the ice grows down into the water, becoming thicker**. You may see curious patterns on top where water could squeeze through the ice.

** Do not venture onto pond ice without an adult to check the safety.

Ice can tell more stories...

Ice can bend and curl.

Hair ice from a log

Compared to needle ice, hair ice is thinner, bendier, and longer.

Ribbon ice

Hair ice grows from the bottom like needle ice, except instead of drawing the water out of the ground, each hair of ice draws water from a plant's pore.

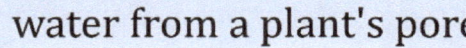

Ribbon ice is like hair ice, but grows from plants with ribbon-shaped pores.

Hair and ribbon ice tend to form in the evening only to melt the next day.

Few people notice them. If you see hair or ribbon ice, you are having a very lucky day!

And ice can sprout as a white fuzz.

Hoar frost is common, often covering large ground areas with a tinge of white ice. The whiteness is from many tiny, densely packed ice crystals.

Hoar frost crystals are like snow crystals: they both grow by collecting vapor. But unlike snow crystals, which form while falling, hoar crystals anchor to surfaces that stick out, such as blades of grass and fence posts.

They tend to form out in the open on a clear evening, only to vanish in the sunshine of the next day.

Hoar frost on grass.   Columnar hoar   Tabular hoar

Neighboring hoar frost crystals look very similar to each other because they form and grow under nearly the same conditions. But, as with snow crystals, they are not exactly alike.

Ice can dribble down...

Icicles start with a small drip that freezes. Then, they grow downward as more water flows down their side, freezing along the side and bottom.

...as well as glob upward.

If water drips off an icicle to colder ground below, it can freeze onto ice that then grows upward from the ground. These upside-down, blobby icicles are called "ice stalagmites".

Icicle tips can be hollow.

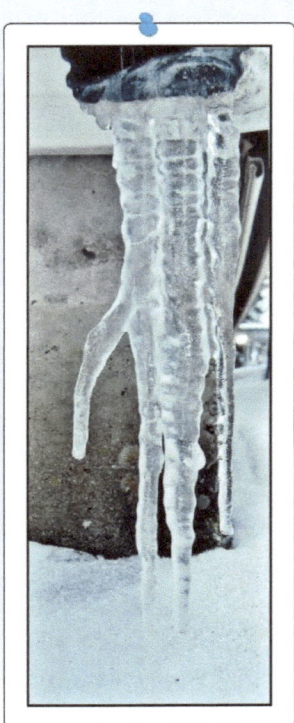

Icicles are usually quite bumpy.

If an icicle is growing, you can usually insert the tip of a toothpick in the end as shown. Otherwise, either the water flow has stopped (too cold), or the icicle is melting (too warm).

Ice can grow into the wind.

The air is below 32 °F (0 °C) and fog has drifted through, leaving white icy deposits.

From afar, the ice looks like hoar frost, but this ice is rime.

Rime comes from supercooled droplets.

Rime ice

Rime builds up on objects in a cold fog where the wind hits. Here, the fog droplets are below the "freezing point" of 32 °F (0 °C), yet not frozen.

These are called "supercooled" droplets.

Many cold fogs have such supercooled droplets. When they touch a frozen surface, they freeze to become rime ice.

Rime grows into the wind.

Rime on lamppost's spider web

Find rime on objects that have cooled below 0 °C. Here, the rime formed on tiny spiderweb strands in the air, but not on the large lamppost in back (which was anchored in warm ground below).

Ice can swirl...

Film frost on car hood.

...and twist.

Film frost on car window.

On cold winter days long ago, the inside surface of house windows could get a thin film of water that would freeze into curvy, beautiful patterns. This is called "film-frost".

Homes these days tend to be too warm to get film-frost. Instead, look for such frost patterns on the outside of cars parked outdoors in freezing weather.

You can find such wonders of nature even in a big city.

Even in summer,

Ice in clouds can make rain.

A rain cloud may look like grey cotton candy, but instead of sugar, it has water droplets and ice. Some of the ice in a rain cloud can look like the pictures on the next page.

When small ice crystals grow larger, they become snow crystals. At left is a dendrite, at right is a column.

Snow crystals also get rime ice. This one is called a rimed columnar crystal.

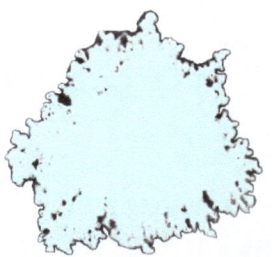

These small pieces of ice show flat crystal faces. They are called "small ice crystals".

Graupel ("graw-pell") forms when the rime completely covers the crystal.

But what is the ice doing up there?

Why is it crucial for rain?

Let's look inside the rain cloud to find out.

Ice grows fast in a cloud...

A droplet in a cloud freezes, and the ice crystal grows larger via several ways. Eventually, it melts into a large drop that may fall on your head. See A) to G) for how it works.

C) Eventually, it freezes and becomes an ice crystal. (Some droplets freeze first because they happen to have a speck inside that causes freezing at a higher temperature.)

D) Water vapor sticks more strongly to ice than water, so the ice crystal grows faster than the nearby water droplets. It grows heavier, yet is still going up.

B) The updraft pushes the droplet higher, where it is colder. It grows bigger.

A) It all starts in a place that has rising air (updraft). Air can rise for various reasons, but rain clouds form only where enough water vapor is rising in an updraft. By rising, the vapor becomes cold enough to stick to a tiny speck in the air. That is how a droplet forms.

...and then can melt into a raindrop.

E) The ice crystal grows even faster when it hits water droplets that stick and freeze. This is riming, just like how rime ice grows. The ice is now too heavy to rise. It starts to come down.

F) Soon, the rimed crystal becomes blobby graupel ("graw pell") roughly a thousand times larger than the droplet it all began with. It falls faster, making it hit more and more water droplets at an ever-faster rate. Hitting and sticking with other ice crystals also makes it grow faster.

G) Falling into warmer air below, it melts into a raindrop.

Ice crystals are often crucial for rain because they help to create large enough drops that can reach the ground before evaporating away.

29

Ice makes thunderstorms...

...electric.

Thunderstorms occur in huge cumulonimbus ("cue mew loh nim bus") clouds—the Godzillas of the cloud world. Compared to smaller rain clouds, their updrafts are faster, extending the cloud upward—often making the cloud top higher than the top of Mt. Everest, the highest mountain in the world.

They also contain more ice as well as larger water droplets than smaller clouds.

With the faster updrafts and greater amounts of water, their icy graupel grows quicker and for a longer time. That way, the graupel can get even bigger and become hail.

Look! A passenger airplane is flying near the top of the thundercloud.

The cloud is indeed very tall.

The updrafts have small ice crystals going up past the large graupel and hail falling down. Their collisions produce electricity.

When a small ice crystal bounces off a larger graupel or hail, it collects positive charge (+) that it then takes to cloud top. Negative charge (−) remains on the graupel or hail and so these negative charges stay lower in the cloud.

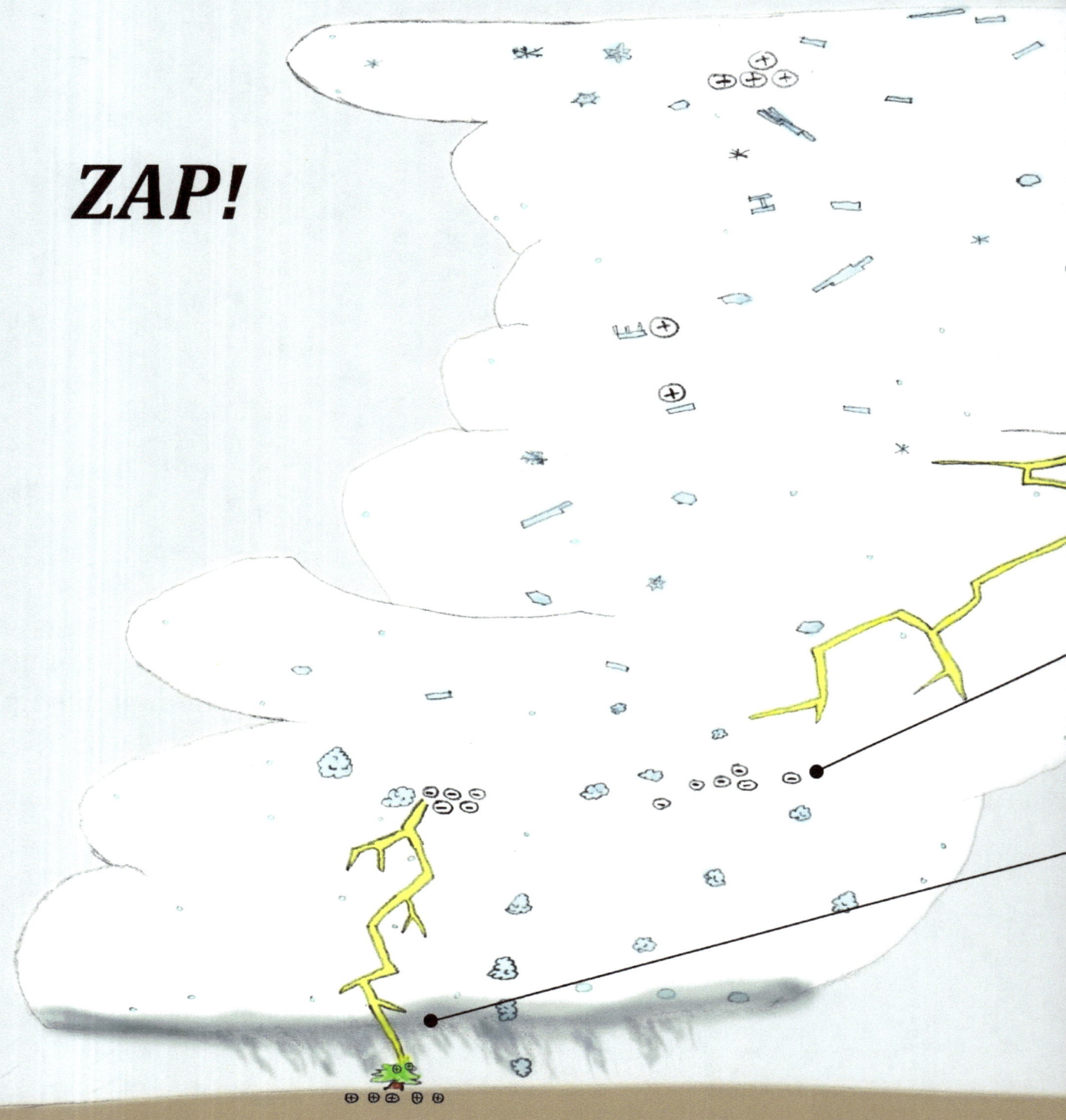

ZAP!

Lightning shoots across the sky.

And to the ground.

The electricity travels in a flash between the negative and the positive regions.

The negative region often lies low enough to the ground that it attracts positive charges in the ground, making them flow to places as close to the cloud as possible—the top of that tree, for example. Then, the nearby air becomes highly electrified, triggering a very dangerous lightning bolt.

Lightning is dangerous because it may strike a nearby object on the ground like a tree, electric pole, or even a person.

Any blobby ice you see larger than about ¼ inch (0.5 cm) across is called hail. The biggest one yet found is about 7 inches (18 cm) across.

About this time, some hail may reach the ground. There, on the ground, lies the main cause of the thunderbolts: the ice.

All year long—
Ice can color the sky.

The colorful "CZ" arc bends up, unlike that of a rainbow.

High, thin clouds dim the sun. If you block the sun from your eyes*, you may see a colored arc above (or below) the sun.

These colors come from sunlight bent by passing through tiny ice crystals that act as a prism.

Seeing an arc tells you that ice is up there. As it is not made by raindrops, it is not a rainbow.

* Make sure that something is blocking your view of the sun. Never, ever look at the sun. Never!!

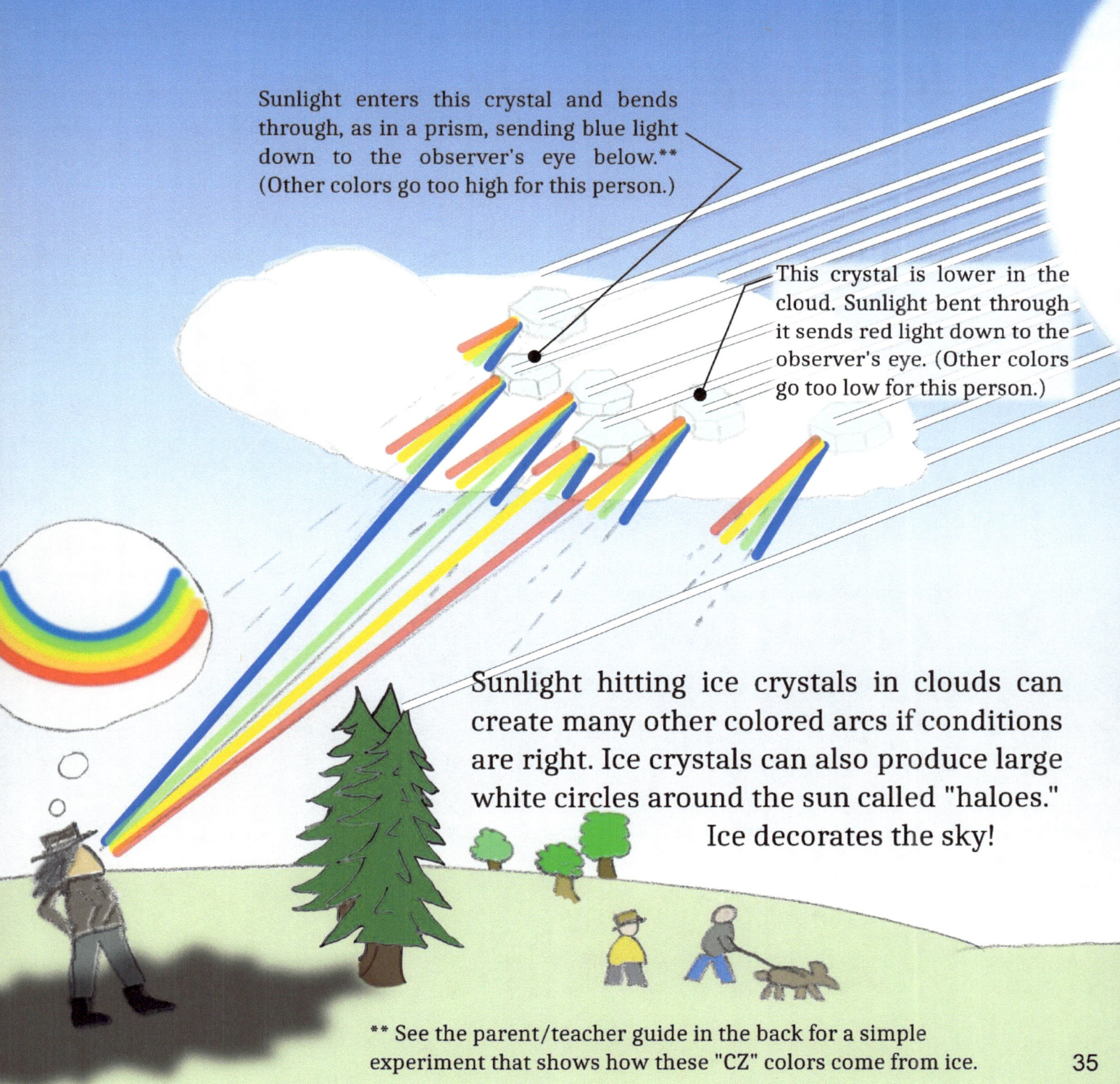

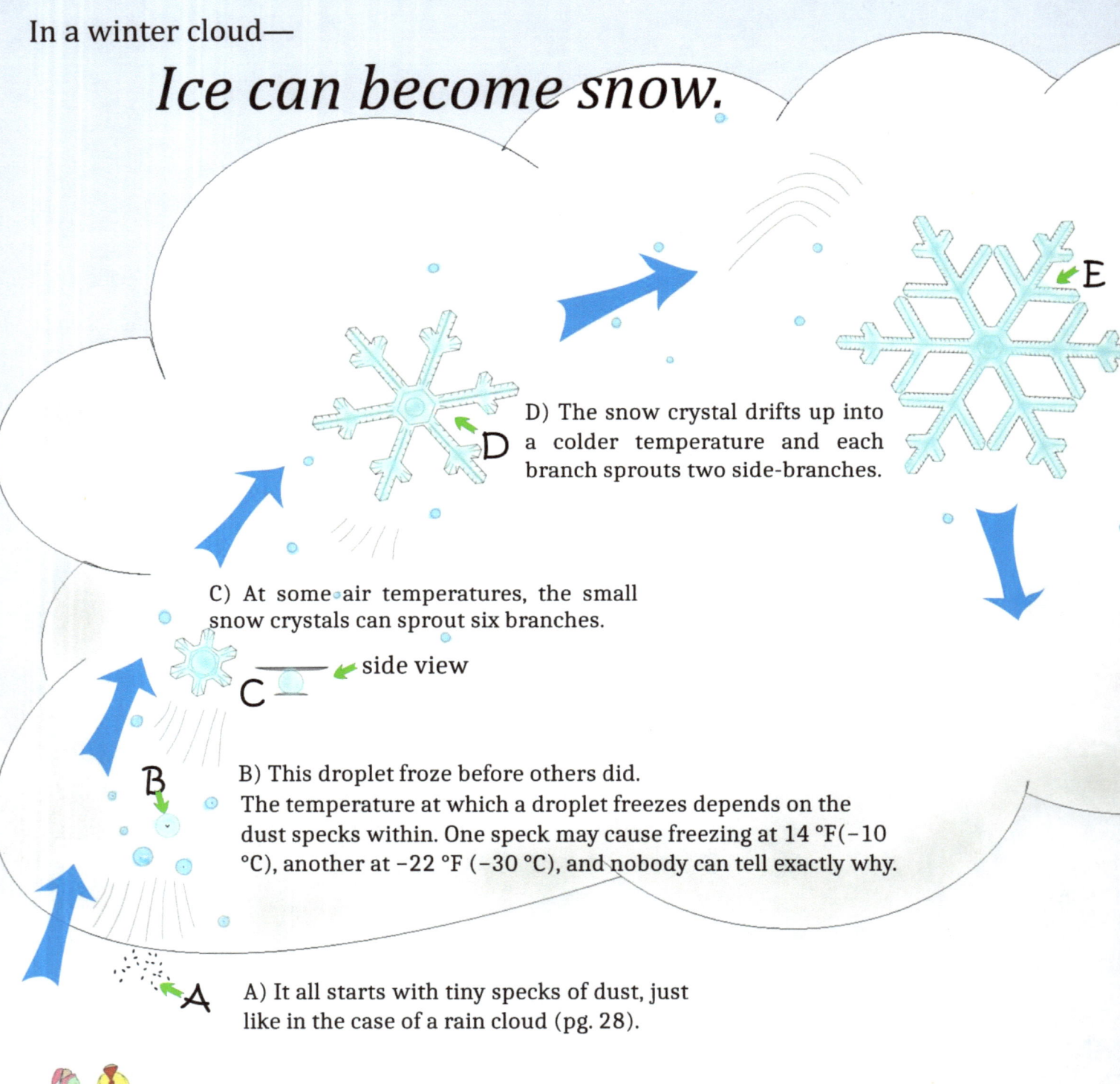

In a winter cloud—
Ice can become snow.

D) The snow crystal drifts up into a colder temperature and each branch sprouts two side-branches.

C) At some air temperatures, the small snow crystals can sprout six branches.

← side view

B) This droplet froze before others did. The temperature at which a droplet freezes depends on the dust specks within. One speck may cause freezing at 14 °F (−10 °C), another at −22 °F (−30 °C), and nobody can tell exactly why.

A) It all starts with tiny specks of dust, just like in the case of a rain cloud (pg. 28).

E) More side-branches sprout, the crystal grows heavier, and falls down.

side view

On this crystal, you can still see the original frozen droplet at the center.

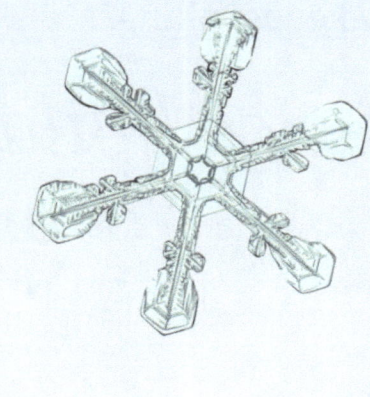

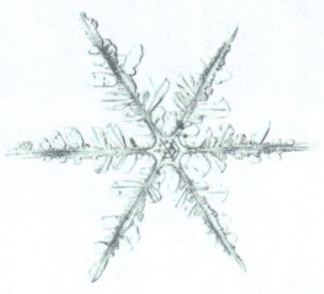

Whenever the snow crystal passes into different air conditions, it grows differently, mainly at its tips.

Many snow crystals grow long and thin like needles. Indeed, they are called "needles".

This crystal hit many droplets, becoming coated with rime (same as rime ice, but on a falling crystal).

An ice crystal in a winter cloud grows like that in a summer cloud. Yet the winter cloud tends to have less vapor, fewer droplets, and slower updrafts. And the air at the ground is colder, allowing the crystal to reach the ground without melting.

During a snowfall, go out and look up at the softly falling snow.

Season's greetings from the sky!

Snow keeps falling...

Ice builds up over many years, pressed into glacier ice.

Lots of air between the fallen snow crystals.

Smaller air bubbles in the firn.

Almost no bubbles in glacier ice.

The weight of snow on top packs the older snow below tighter as the air is squished out. This older, more compact snow is called firn. The deeper the firn, the tighter it is pressed. At some depth, it becomes glacier ice, where the air bubbles are tiny and far apart.

In the coldest polar glaciers, the deepest firn may be about 300 feet (91 m) down and over 2000 years old. The glacier ice below is hard and clear, yet bluish.

Yet deep within, ice retains stories of our past.

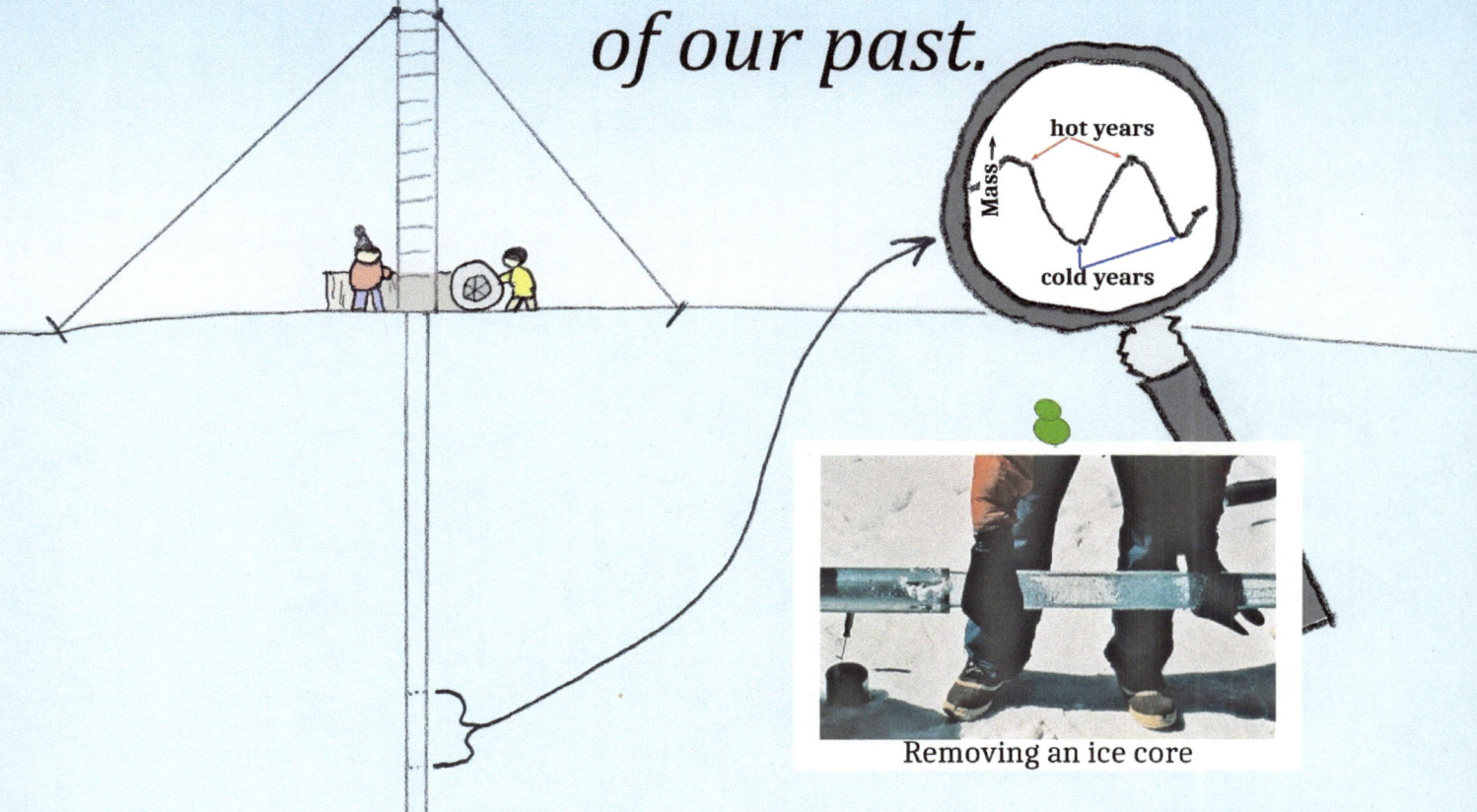

Removing an ice core

Scientists remove long core-sections of ice from glaciers in Antarctica and Greenland. By analyzing the ice, they learn about the Earth of long ago: tiny changes in the mass of the ice's water molecules reveal past air temperatures. Also, chemicals and dust trapped in the ice reveal past conditions such as winds, fires, and volcanic eruptions.

Ice can flow and slide.

Like water in a river, the glacier ice flows down the mountain, as well as sliding along the bottom. But the glacier's ice moves much slower than a river's water—sometimes less than 10 feet (3.0 m) per year.

Ice breaks off and floats away.

After reaching the sea, the glacier breaks up into icebergs that float. In polar regions, some of the glacier ice at this point may be thousands of years old.

The ice, now iceberg, floats off to melt.

Icebergs are much larger than they look, with about 90% of their ice underwater. So, it can take years for the whole iceberg to melt.

Why does ice float?

Ice floats because it is hexagonal.

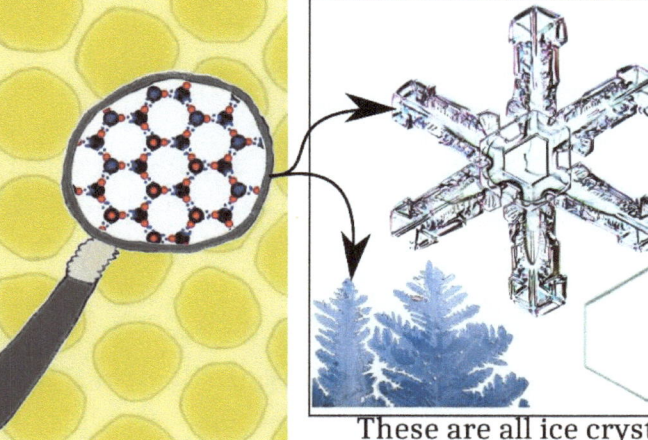

These are all ice crystals.

If you could look deep inside ice, you would see that every crystal has the same exact pattern of water molecules —a hexagonal structure that looks like a honeycomb.

This structure has a lot of empty space. Whereas bees can use it to store honey, ice just has the empty space, making it lighter than liquid water.

So, thanks to the hexagons—

Ice floats!

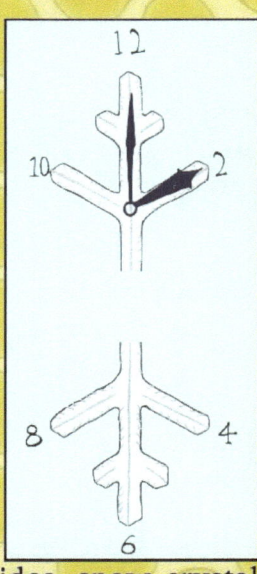

As the hexagonal cells have six equal sides, snow crystal branches often point in six directions. If you line up one branch to 12 on a clock, the other branches can point only to the 2, 4, 6, 8, and 10. Even on the little side-branches.

Also, snow crystals can show six equal sides!

Ice can be both sticky and slippery.

This is strange, no?

You can make snowballs because ice is sticky. Wet snow is the stickiest, but even colder and drier snow sticks together.

Compare it to sand. Can you make "sandballs" out of dry sand?

Ice is also slippery. Think of other big, flat materials you stand on. How well would your ice skates glide on concrete? Not well at all, right? (Please don't try this!)   And yet you can glide far on ice.

So, ice can not only flow and float, but it can also stick and slide. It is indeed quite special.

If Earth never had ice, what then?

Then, we'd have had no ice ages, maybe no people and animals, and certainly no ice cream.

What would our Earth be like without ice? The whiteness of ice makes the Earth more reflective, making the ice ages colder and longer. Over long times, the movements of immense glaciers greatly altered the landscape and likely promoted the evolution of many species, including us. Though we don't know exactly what Earth would be like, things would indeed be very different without ice.

Every glacier, every heavy rain, and every thunderstorm comes from that same crystal that shows us rainbow-like arcs in the sky, gives us cool ice cream, and dazzles us with myriad beautiful forms.

*For all this and much more,
we owe to the little crystal called ice.*

*Next time you see some ice,
take a minute to look closer—*

it has a story to tell.

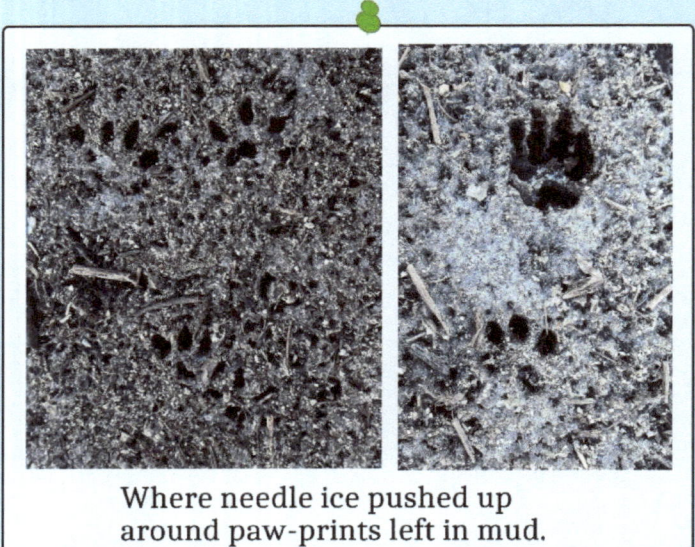

Where needle ice pushed up around paw-prints left in mud.

Tabular hoar frost forms at about 12 °F (-11 °C) or colder.

Fun Facts

- The modern style of ice cream was developed in 1846 by Nancy Johnson, an American woman who first invented the now-classic hand-cranked ice cream maker.

- During WW2, when Americans were asked what they would do first after the war, many spoke of gorging on ice cream. American aviators in Britain who shared the same cravings became very creative. They put the ingredients (cream, sugar, salt, flavors) in a can that was placed in the rear of their aircraft. Then on their mission over enemy territory, the plane's shaking and the high-altitude coldness froze the can's contents to ice cream.

- Ice nearly 3 million years old has been found in an ice core, from snow that fell before our most recent ice age.

- A layer of hoar frost can form within a snowpack. It is a major concern in the mountains because it makes a snow slope unstable and more likely to cause an avalanche.

- One of the most influential researchers of ice and snow was a self-educated farmer in Jericho, Vermont. Wilson A. Bentley (1865-1931) was well-recognized and respected outside of his hometown, but his father, brother, and other residents of Jericho all thought he was "wasting his time" in his studies.

- Glaciers typically move quite slowly. But occasionally, one can suddenly speed up for several months, advancing over 300 feet (90 meters) in a single day.

- Currently, scientists recognize about 87 types of snow crystals, each having a distinctive shape.

- The iceberg that sank the Titanic in 1912 was estimated to be 50-100 feet (15-30 m) above water, meaning that it may have been up to about 900 feet (270 m) deep.

- Alfred Wegener, famous for his theory of continental drift, also was the first to explain why ice grows rapidly in clouds to later melt into raindrops. He got the idea after observing ice and droplets on window glass around 1910. But as far back as the late-1700s, scientists (including Benjamin Franklin) thought that summer rains may often originate as melted snow.

- All natural ice observed on Earth has the same hexagonal-type structure inside (pg. 43). But in the laboratory, researchers have put water under various extreme conditions that result in at least 20 other inside ice structures. This number for ice is more than that from any other known crystal.

Parent/teacher-guided activities

These experiments demonstrate principles in physics, chemistry, and weather. To achieve success, you may have to play around and vary the "recipe" according to the particular "ingredients" you have handy. If understanding eludes you at first, don't panic, for it happens to all of us and indeed ultimately may be helpful. The activity itself is the important thing. You can't fail to learn from it.

Experiments to do at home

I) Noisy ice (physics). Why do some ice cubes crack when placed in water?
Needed: ice cubes, fresh out of a freezer, two glasses of water.

Remove a few ice cubes from the freezer and let them sit for a few minutes. At this point, their surface should appear wet and their interiors should have warmed up to nearly the melting temperature of 32 °F (0 °C). (In the freezer, the ice cubes' temperature was likely near 5 °F, or -15 °C.) Drop them in one glass of water. Notice that they don't crack. Next, remove a few more ice cubes from the freezer and immediately drop them in the second glass of water. In this case, you may hear a sharp snap and notice that they have cracked. These ice cubes cracked because the contact with water rapidly warmed their outside to 32 °F (0 °C) while their interior remained at the freezer temperature of about 5 °F (-15 °C). Ice, like most materials, expands when warmed up. So, the outside expanded while the inside did not, thus creating cracks and making cracking sounds.

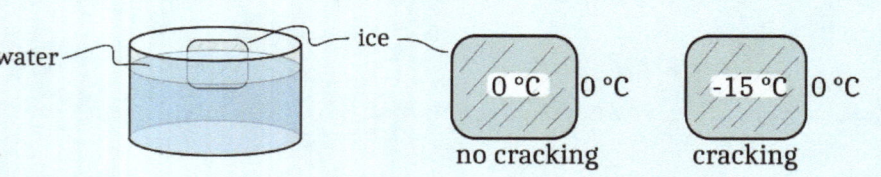

| Temperature conversion ||
°C	°F
20	68
10	50
0	32
-5	23
-15	5

If it doesn't work, try using larger or colder ice cubes.

II) Colorful ice (physics, weather). What causes the rainbow-like CZ* arc, as shown on page 34?
Needed: A tall, smooth-sided clear container and a windowsill in the sun. A good container to use is a clear plastic two-liter soda jug, cut-off near the top. You can also use a tall, straight-sided drinking glass, but the cut-off soda jug is useful for other experiments and gives a larger arc of colors. (We use water in this experiment because it is easier to use, but the result is the same as that with clear ice.)

* Stands for "circumzenithal arc", meaning that it is a segment of circle around the zenith (straight up).

Activity is the only road to knowledge.
— George Bernard Shaw

Now fill the jug nearly to the top with water and place on the windowsill, near the edge. Look for a rainbow-like arc on the floor in the dark shadow of the wall. Put a white sheet of paper on the floor to make the arc clearer. If you fail to produce the arc, try again when the sun is in a different position. The best sun angles are between about 20° and 45° above the horizon. To see how this can explain the CV arc, study the diagrams below and on page 35.

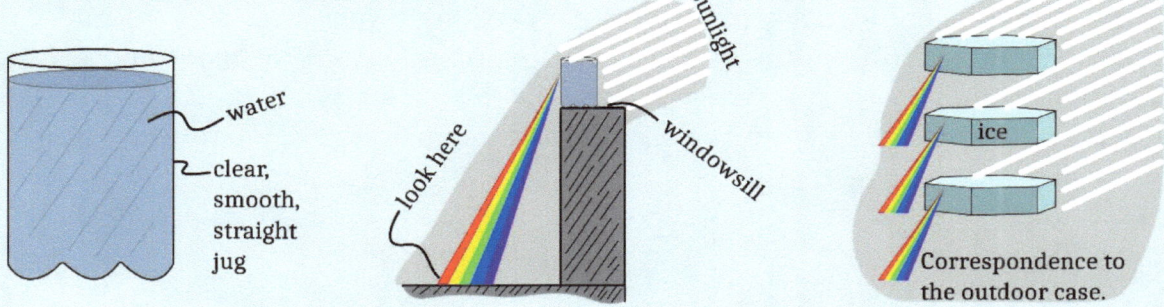

The bending of the sunlight in water is refraction, and the spreading out of the refracted white light into colors is dispersion. Find dispersion in other clear materials including glass and acrylic. The remarkable thing about the CZ arc is that the crystals must be almost perfectly horizontal as they fall.

III) Supercooled drops (chemistry, weather). Can you supercool* water?
Needed: Water, lid to a jar, oil or vaseline, the jar (or a plastic cover) for the lid, and a level space in a freezer to set the lid and cover. Refer to experiment I for temperature conversions.

Smear a thin layer of vaseline or cooking oil on the inside of the lid (bottom, here). Then place about six small water drops onto the same surface. Due to the oily surface, the drops should bulge out like a hemisphere. Put the cover over the bottom (lid) to protect the drops, and put it all into the freezer. After 15 minutes, remove from the freezer and take off the cover. If any drops froze, they will look whiter than the others, whereas the supercooled ones will jiggle when tapped. If your freezer is below -15 °C, then eventually all drops may freeze, yet you should find supercooled drops by removing the lid within 15-30 minutes.

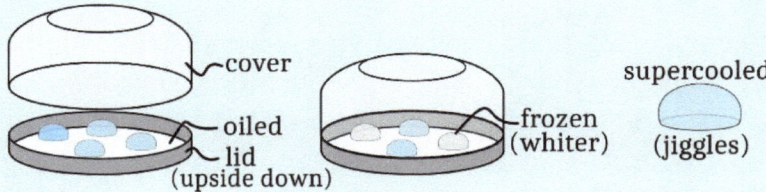

* See pages 23, 28, 36 for the meaning and importance of supercooling.

IV) Heat in ice (chemistry, physics). When melting, why does the ice temperature stay near 0 °C (32 °C)? Needed: Two large cups, ice cubes, water and a thermometer.

This experiment has two parts. Part 1: Fill one cup (a) less than halfway with room-temperature tap water and measure its temperature (say, 20 °C here). In the other cup (b), prepare the same amount with roughly 0 °C water by adding an ice cube to about 1/2 cup of water and stirring until the cube melts. (See table below for all temperature conversions.) Now pour cup (a) into cup (b), stir, and remeasure the temperature of the mixture (c). Note that the value is about midway between the two original temperatures. For part 2, repeat 1 except for (b') use ice of 0 °C (i.e., as in experiment I, wait until the ice visibly starts melting). Note how the resulting temperature of mixture (c') remains near 0 °C.

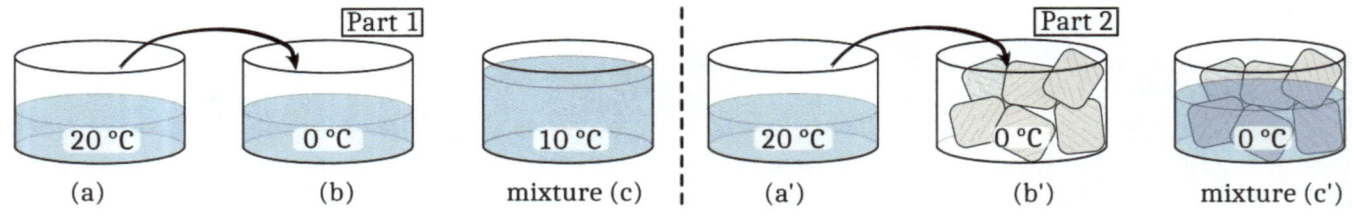

In part 1, thermal energy as heat flows from (a) to (b), thus cooling the water of (a) and warming that of (b). The resulting temperature is midway in-between. In part 2, the heat in (a') is used instead to release the water molecules from the ice structure in (b') without warming, that is, to melt the ice. So, the temperature of mixture (c') does not rise above 0 °C until all ice melts. (The reverse occurs during freezing: heat produced by freezing warms the ice and water to about 0 °C until all the water is frozen.) See table at right for all temperature conversions.

Temperature conversion

°C	°F
20	68
10	50
0	32
-5	23
-15	5

V) Melting ice with salt (chemistry). Why does salt melt ice?
Needed: About 20-30 ice cubes in a large bowl, about 1/2 cup of salt, a thermometer, and something to stir the ice in the bowl.

Add the salt to the bowl of ice, wait a few minutes for some ice to melt, and start stirring. Once the salty melt-water is deep enough, use your thermometer to measure the temperature. You may find that the temperature is below -15 °C. To convince yourself that the salt-water-ice mixture is really that cold, make a second bowl with ice and water (no salt) to compare. Try sticking your finger in both bowls. The one with salt should indeed feel much colder, in agreement with the thermometer.

> *We see nothing truly till we understand it.*
> — John Constable

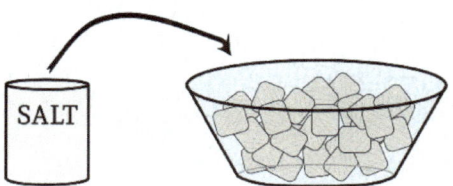

Why? To help understand, picture the boundary between ice and water as a region that, though looking steady and unchanging on the large scale is actually a region of great activity on the microscopic scale. Here, molecules are constantly detaching from the ice and merging into the water, while water molecules are constantly striking the ice surface and sticking onto the ice. If the ice is neither melting nor growing, then the number of molecules leaving the ice equals the number of water molecules sticking to the ice. A balance exists. But if the ice is melting, more molecules leave the ice than stick. Salt in the water prevents some water from sticking, yet the number leaving the ice surface is unaffected by the salt. So, a net amount of melting proceeds despite the temperature being below 0 °C.

The reason why the mixture cools to near -15 °C is the same as that in the melting in experiment IV above: the melting ice draws heat from the surrounding liquid, cooling the whole salty water-ice mixture.

VI) Freeze patterns (physics, chemistry). How does water freeze?
Needed: A clear soda jug, cut out as explained in experiment II, ice cubes, and a place that is colder than about -5 °C for at least several hours per day.

Put water plus some ice cubes into the jug, wait until the ice cubes just finish melting, then move the jug outside, preferably in the shade. Then simply wait for the water to freeze. If it is quite cold outside, you might need to wait only a few minutes before ice starts forming. Otherwise, check every twenty minutes or so to observe the freezing. Results are faster at colder temperatures. After much of the top surface freezes, try reaching in and removing some ice to observe the ice close-up. To make it easier to remove, warm the container with your hands, then carefully turn over the jug to remove the ice in one piece, allowing the remaining unfrozen water to drain (to the ground, not your lap!) Every time you do this, the freezing pattern will be different, though you will note some similarities. You can even use several jugs at the same time to see how the freezing pattern can vary even under the same conditions.

start

frozen top

turned upside down, drained, & ice removed

Outdoor observations

This book's story introduces many outdoor observations of ice you may make. Go outside on cold winter days and find these as well as other forms. Some are shown on the back cover, yet endless combinations and variety exist. All you need for discovery is to remain curious. Bring a camera to record what you see. An enlargement may reveal unexpected features.

For example, you might first come across a freezing puddle. If the puddle is not completely covered by ice, try breaking off a piece growing towards the center. (It helps to wear snug rubber gloves to hold the ice and keep your fingers warmer.) If the conditions are just right, you may be lucky and find a piece of ice growing in the middle of a large puddle, or in a pond, unattached to the shore. Very carefully lift this ice out (e.g., using small sticks) and compare it to the "melt-grown dendrite" on the back cover.

Straight and curved lines near a pond's shoreline.

A large single ice crystal from the shore of a slow-moving stream.

Other things to look for:

- Ice-covered puddles: Examine the underside and try to guess the cause of the various lines.
- Rime: If the rime is relatively clear, the drops froze slowly due to being larger or warmer.
- Hair ice: Look in wet, soggy areas that experienced rapid cooling. The hair's "parting line" shows where the surface was coldest.
- Hailstones: Break them in half and notice the layers showing warmer and colder growth.
- Snow on the ground: Look for weird textures from wind or melting.
- Arcs: Many sky patterns arise from ice in thin clouds and fogs. Learn of them and appreciate the presence of ice year-round.

There is no surer road to fairyland than that which leads to the observation of snow forms. To such a student, the winter storm is no longer a gloomy phenomenon to be dreaded. Even a blizzard becomes a source of keenest enjoyment and satisfaction, as it brings to him, from the dark, surging ocean of clouds, forms that thrill his eager soul with pleasure.

— Wilson A. Bentley

*The snow is melting
the village is flooded
with children*

— Kobayashi Issa

The deeper one enters into the study of Nature, the further one ventures into and along the by-paths that, like a mystic maze, thread Nature's realm in every direction, the broader and grander becomes the vista opened up to the view.

— Wilson A. Bentley

About the authors

Jon (left) got his physics PhD at the University of Washington studying ice in clouds. He continued his research, first running the cloud physics laboratory at the University of Arizona, and later at a university near Kyoto, Japan. His research helped reveal how ice and snow crystals grow as well as how ice crystals charge up thunderstorms. In 2009, he co-authored The Story of Snow (Chronicle Books).

Growing up in Tokyo, Japan, Sam (right) experienced winter snow only as big and wet snowflakes until moving to Colorado. Here, she was captivated by the thin, star-shaped snow crystals slowly drifting down. An important writing advisor to Jon's various manuscripts, as well as an award-winning story-writer in Japanese, she is happy to make The Story of Ice her first official book in English.

We began this book project soon after The Story of Snow came out, revising many drafts over the years. Has it inspired you to go outside and see new things or perhaps see old things with new eyes? If so, please let us know.

www.ingramcontent.com/pod-product-compliance
Lightning Source LLC
Chambersburg PA
CBHW061158030426
42337CB00002B/43